Hydroponic Gardening Secrets:

The Complete Beginners Guide to Learn How to Start Hydroponics from Scratch. Perfect Hydroponic System to Grow Your Fruit, Vegetable and Herbs.

© Copyright 2020 by Viktor Baras. All right reserved.

The work contained herein has been produced with the intent to provide relevant knowledge and information on the topic on the topic described in the title for entertainment purposes only. While the author has gone to every extent to furnish up to date and true information, no claims can be made as to its accuracy or validity as the author has made no claims to be an expert on this topic. Notwithstanding, the reader is asked to do their own research and consult any subject matter experts they deem necessary to ensure the quality and accuracy of the material presented herein.

This statement is legally binding as deemed by the Committee of Publishers Association and the American Bar Association for the territory of the United States. Other jurisdictions may apply their own legal statutes. Any reproduction, transmission, or copying of this material contained in this work without the express written consent of the copyright holder shall be deemed as a copyright violation as per the current legislation in force on the date of publishing and subsequent time thereafter. All additional works derived from this material may be claimed by the holder of this copyright.

The data, depictions, events, descriptions, and all other information forthwith are considered to be true, fair, and accurate unless the work is expressly described as a work of fiction. Regardless of the nature of this work, the Publisher is exempt from any responsibility of actions taken by the reader in conjunction with this work. The Publisher acknowledges that the reader acts of their own accord and releases the author and Publisher of any responsibility for the observance of tips, advice, counsel, strategies, and techniques that may be offered in this volume.

Table of Contents

Introduction..6

Chapter 1: A Brief Overview of Hydroponics..9

Chapter 2: Design and Plant Your Dream System...................................18

Chapter 3: The Constant Gardener..39

Chapter 4: Maintaining Your System..59

Chapter 5: Next-Level Hydroponics..71

Conclusion..88

Introduction

Welcome to *Gardening Secrets*, your all-in-one guide for designing, assembling, and growing the hydroponics garden of your dreams. Gardening has historically been a way for people to connect to their food and beautify their surroundings, provide fresh produce for their families, and as a healthy, productive hobby. But what if you have no place to put a traditional garden?

Hydroponics- the practice of growing plants without soil- can help you bring the outdoors inside with some simple equipment and a little know-how, and you've come to the right place to find it! If you have ever given thought to indoor growing, but you've never been sure where or how to get started, the comprehensive information you'll find here will help you decide if hydroponics is right for you and choose a budget and a style that's in line with your needs.

Aside from space considerations, people grow hydroponic gardens for many other reasons, including mobility issues, a desire to add to an existing outdoor garden, local ordinances that limit the size of traditional gardens, or an innate curiosity about the practice. For those interested in horticulture, it can become an expansive hobby- hydroponics leaves a lot of leeway for experimenting with different nutrient mixes and growing different varieties and cultivars.

In this book, we'll go over a quick history of hydroponics, and give an overview of the different growing techniques. You'll also get the advice you need to help you choose the right hydroponics system to fit any space or budget, and give you the inspiration to design and build a garden that is user-friendly, economical, *and* aesthetically-pleasing. We'll also address what type of supplies you'll need to get your garden underway, and you'll learn about seed selection, basic botany, how hydroponic systems can be adapted to meet the needs of a variety of plant species, and how to get your system up, running, and planted.

You'll also learn about the best ways to facilitate cleaning and maintaining your system, as well as how to schedule succession planting so that you will always have something green, growing, and ready to harvest for your enjoyment. We will explore ways that you can get the most out of your hydroponics system, including water testing, fertilizer applications, and adjusting your nutrient mixes and growing mediums. You'll also learn about some techniques for creating a garden journal to keep track of your planning, planting, and harvesting activities.

Each chapter also contains helpful lists and recaps to help you stay organized as your progress through choosing, designing, building, planting, and maintaining your system.

It should be noted, no statements in this book should be considered to be medical advice, and readers should consult their health providers should they have any questions regarding the nutritional or medicinal benefits of hydroponically grown food and other plant matter. While an increasing amount of commercially available food may be produced hydroponically, a physician, internist, or dietician can help you make decisions about the benefits of home-grown hydroponic food.

Gardening should be fun, productive, and provide a boost to your mood and your health- and a solid hydroponics system can help you do that right in the comfort of your own home. If you think hydroponics is right for you and you're ready to take the plunge into the world of soilless growing, *Gardening Secrets* is here to take you from the first step of planning through a bountiful harvest, so let's get started!

Chapter 1: A Brief Overview of Hydroponics

Hydroponics is becoming increasing popular world-wide as a method of hobby gardening and mass food production. One of the vast advantages of hydroponic growing is that there is very little waste in a well-designed hydroponic system, and as technology advances, these hydroponic systems also become more efficient.

A Quick History of Hydroponics

While modern hydroponics has only been in development and use since the 1900s, there is some evidence that the famed Hanging Gardens of Babylon may have featured hydroponics in 600 B.C. Sir Francis Bacon, the father of the scientific method, also explored hydroponic growing in the 1600s, with extensive descriptions of his practices published posthumously in the 1627 book *Sylva Sylvarum*.

Ancient wonders of the world aside, hydroponics, as we recognize them today, got their start when University of California at Berkeley professor William Frederick Gericke began experimenting with what he called aquaculture (unaware that the term was already in use for another discipline) in the 1920s; he would later coin the term hydroponics. Gericke would become famous for growing tomatoes, some allegedly being cultivated on vines nearly three stories tall, using a soilless method of gardening.

Gericke's experiments were nearly debunked when he refused to share his research results with other agriculturalists, and the resulting fallout led to him leaving his post at the university. In 1940, he would publish *The Complete Guide to Soilless Gardening*, revealing his methods and techniques to the public for the first time. Gericke's torch was taken up by a botanist named Harold Resh, who would publish his own guidebook in 1978. The mid-to-late 1900s also saw extensive hydroponics research done by NASA, in the pursuit of techniques which could be used for extraterrestrial food growth.

In the last few decades, technological advances have changed the face of hydroponics once again, but all the basic principles remain the same as when Gericke began his work in the 1920s. Hydroponic cultivation aims to remove the uncertainty of soil quality from the equation of plant growth and replace that soil with carefully controlled airflow, water, and nutrients. Hydroponics also aims to fulfill the exact needs of each plant to promote strong, healthy, measured growth. Let's take a look at those needs and how hydroponics works to provide them.

Basic Botany as Applied to Hydroponics

In order to grow properly, plants, like all living things, need a handful of requirements to be met. In this section, we'll dive into some basic botany and go over how those basic needs can be met with hydroponics instead of through traditional growing methods.

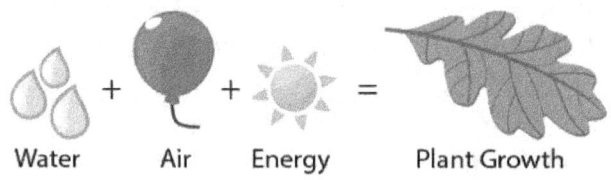

- all plants need their basic requirements met to grow properly-

Water- Plants need water to grow, but not all plants need the same amount of water. A cactus does not require watering as often as an azalea. The beauty of a hydroponics system is that each plant can take up as much water as it needs, and no more. The rest of the water is recycled back through the system for reuse. Hydroponics can actually use much less water than traditional planting because the water stays in the system instead of being leached out through the soil.

Water is vital to proper plant growth and sustenance as it helps carry nutrients throughout the plant's circulatory system, which generally consists of the xylem, phloem, and capillaries. These structures carry water and nutrients through the stem and leaves of the plant, and not only is water the conduit for transport, but it also helps hold the structure of the plant, known as turgor. Without enough water, a plant will lose turgor and show signs of wilting and slumping.

Air- Proper airflow is important to plants for more than one reason. Plants need air to respirate and photosynthesize, but good airflow is also vital to warding off pests and plant pathogens. Plants don't 'breathe' as humans do, but they do need atmospheric CO_2 (carbon dioxide) to fuel their photosynthesis and O_2 (oxygen) to fuel respiration. Plants use these atmospheric gases to perform these functions, utilizing CO_2 and water to create food, and utilizing O_2 and water to move that food around the plant and break it down into energy. The plant then releases the excess O_2 back into the atmosphere.

In a hydroponics system, atmospheric airflow is also important for plant health because the growing technique is, by its very definition, wet. Moisture can invite plant pests and disease, so it's crucial to give your plants enough space not to allow insects and diseases to move in and take hold. Although removing soil from the equation takes away the risk of soil pathogens, you still want to protect your plants from insects and infections.

Air temperature is another way that the atmosphere affects plants. Although it's highly likely that your hydroponics system is inside and protected, you still need to make sure it's not in an area of your home that's not too hot or too cold. Just as plants behave outdoors, so will they behave in a hydroponics system if they are not kept at an amenable temperature indoors.

Light- Without light, a plant cannot perform photosynthesis, which is the biological process of creating food for itself to grow and produce flowers or fruit. Photosynthesis is from the Greek 'photo' meaning light, and 'synthesis' meaning to put together. In all green plants, specialized cells that are known as chlorophyll take CO_2 and water, powered by light energy, and make glucose to feed the plant's functions. The chemical formula for this is:

$$6CO_2 + 6H_2O + \text{light energy} = C_6H_{12}O_6 + 6O_2$$

The carbon dioxide and water plus the energy from light combine to make sugar for the plant to consume, and the remaining oxygen is then released from the plant. You can think of your plant's green surfaces, like leaves, as little solar panels designed to take up the energy from the sun or an alternate light source. If you are going to be setting up a hydroponics system, you may need to provide this light in the form of artificial grow lamps, which are set to emit the proper ultraviolet frequency needed for proper plant growth.

Nutrients- In a traditional growing system, the nutrients for your plants would be in or added to the soil, but because a hydroponics system does not have soil, you'll need to provide these nutrients in another way. Plants need a few basic macronutrients to grow properly, as well as micronutrients to aid in certain functions. When you have a hydroponics system, you will be providing these nutrients via soluble minerals in your water supply.

The most important nutrients for plants are nitrogen, phosphorus, and potassium (N, P, and K). Nitrogen is a component in chlorophyll, and without well-structured chlorophyll, a plant cannot feed itself.

Nitrogen also contributes to the building of amino acids, helping the plant grow on a cellular level. Plants who are lacking in nitrogen will exhibit yellowing and leaf curl, or excessive leaf drop.

Phosphorus is vital for plant growth because it facilitates almost every biological function in the plant. Phosphorus aids in the movement of food, the building of cells, and the formation of genetic material. Without phosphorus, plant growth will be stunted, and plants may not be able to reproduce; symptoms of phosphorus deficiency will manifest in small, weak seedlings and fruit that grows with small or malformed seeds.

Potassium in plants, like in humans, is an electrolyte. It is responsible for the activation of enzymes necessary in both the respiration process and photosynthesis, and, like phosphorus, it has a role in almost every operation of the plant. When a plant does not receive enough potassium, it will suffer from browning, yellowing, and leaf curl, and may not be able to produce fruit or flowers.

Plants also require micronutrients, which will contribute to healthy specimens, but in much smaller amounts. These micronutrients include but are not limited to, iron, magnesium, calcium, sulfur, manganese, copper, cobalt, and zinc. In traditional gardens, these elements are found in the soil, and so hydroponics gardeners will have to supply them to their plants in their soluble nutrient mix.

Soil- But wait a minute! We're removing soil from the equation, so why are we listing it here? Because soil is important for plant health for more than just providing nutrients. Soil gives a plant something to hold on to so that it can grow upright and strong. If a plant is not growing in soil, what will its roots be based in so that it can use its energy to grow and not hang on for dear life?

In lieu of soil, you'll have to give your plants a growing medium, and there are many to choose from. Some hydroponics gardeners like to use mineral components like perlite or vermiculite, and some like to use softer materials like coir, which is made up of the fibers found on the hull of the coconut. When you design your system, you will have to decide what kind of growing medium you'd like to give your plants, so that their roots have structure and security.

Chapter 1 Recap

In this chapter, we went over some important points to remember moving forward:

- hydroponics has been used for centuries but was modernized in the early- to mid-1900s

- hydroponic systems are designed to eliminate the need for soil in plant growth

- all plants need their basic requirements met to thrive:

> - water to carry nutrients and maintain cell structure
>
> - air for respiration and airflow for plant health
>
> - light for growth and photosynthesis
>
> - nutrients to nourish the plant
>
> - soil or growing media for roots to take hold and keep plants upright

Now that you've got a good idea of what plants need to grow and how a hydroponics system can work to fulfill those requirements, let's begin to think about what kind oy system you want to use. In the next chapter, we'll be going over the basic types of systems, their commonalities and differences, and what supplies you will need to design and build a system that works for you, your skills, and your budget.

Chapter 2: Design and Plant Your Dream System

Not all hydroponics systems are created equal, and it's important to understand the components that go into building a system so that you can design one that works within your space, skills, and budget. While there are certain building blocks common to almost every hydroponics system, they can be adjusted and customized to fit just about any need.

Types of Systems and Techniques

All hydroponics systems function on the same basic premise- that by taking soil out of the growing equation and replacing it with a customized nutrient solution, the guesswork, and uncertainty of soil health is eliminated. This is a major advantage of hydroponics because soil is a finicky medium and is subject to many outside stimuli that can change its makeup in a very short period of time. Another major advantage to hydroponics, which we already mentioned briefly, is a reduction in waste. Water in the system is recycled, whereas much of the water used in a traditional garden is lost to the soil, only to be taken back up when the water cycle renews.

There are six different basic hydroponics designs, and you can choose what works best for you. If you go to a large-scale hydrofarm or research facility, you might tour through greenhouses that are full of large water tables with the plants sitting in what looks like shallow moving pools. These tables are called flood tables, and they are part of a technique known as **ebb and flow**. These flood tables can be made or purchased in a variety of sizes to fit any space.

Ebb and flow systems pump a water and nutrient solution into a pool beneath plants that are set in trays holding the growing medium. The plants' roots are bathed in the solution, and the water usually flows in one end of the table and is drained back out the other side in a continuous cycle. These systems are very efficient in not wasting any water. Grow lights are usually suspended above the tables if natural light is not available. The ebb and flow system is easily adaptable to any size and is well-suited to growing vegetables, fruits, and herbs.

Another style of hydroponics system that is very popular is the **wick method.** This is the simplest of all hydroponic techniques and can be done without the use of a water pump, although one can certainly be used to avoid stagnation. It works by placing your plants in your growing medium and giving it a wick connecting the plant receptacle to a reservoir of water and nutrient solution. The liquid is drawn up the wick into the plant's roots as needed. While this system is great for starting large plants and maintaining small plants, it's not ideal for heavy feeders or large plants over the long term, because the plant's needs may outweigh the reservoir or the wick's capacities.

The wick method is popular with gardeners who want to experiment with small batches of plant varieties or test out new nutrient solutions but does not work very well on a large scale. It's also frequently used in classrooms as a teaching aid or student experiment. The wick method is great for trying out new things, but as a sustainable technique for hydroponics, you may want to consider what you plan on growing and decide if you have the space or patience for maintaining a wick method garden. A wick system can be placed in a naturally sunny location or supplemented with a grow light.

Another simple technique for hydroponic growing is called **water culture**. This method utilizes a closed tank into which water and nutrient solution is pumped. A bubbler or water stone is usually placed at the bottom of the tank to keep the solution moving. Plants are placed in growing medium and floated on a tray at the top of the tank;

these trays are often made of Styrofoam or other lightweight material. Grow lamps are suspended above the apparatus if natural light is unavailable.

Water culture systems can be made from repurposed aquariums or other watertight containers, and the floats can be sourced from recycled materials as well, to keep costs down. These systems can be as large as the containers you can source, or as small as a dishpan. These systems are great for classrooms or students, because of their low cost and the ability to build them with very little equipment. The downside to water culture systems is that they cannot accommodate large plants, but they are a good cost-effective way to start seedlings for transplant to other systems or into traditional gardens.

The next hydroponics method you may want to consider is the **nutrient film technique** or **NFT**. NFT is what you may think of when you think of a 'traditional' hydroponics system, where the liquid is pumped through pipes in which the plants are in baskets placed in holes along the pipe. The water and nutrient solution is piped along the roots of the plants and back into a reservoir to be recycled.

This system is very efficient and does not waste much, if any, water. It also saves on the cost of growing medium, which needs to be changed in between growing seasons in other systems. The decided downside to this type of system is that a high-quality timer and pump must be used; if there are any power failures, the roots will dry out quite quickly without the protection of a growing medium. However, an NFT system can be built or purchased in just about any shape, size, or strength of structure to accommodate a large variety of plants. Grow lights are added to the set-up if natural light is not an option based on space or design.

One of the more hi-tech options for hydroponic growing is the **aeroponic** design. These systems do not use growing medium, but instead, suspend plants with bare roots over a misting system. These are wonderful systems on a small scale but are not the most cost-effective. They run on a sensitive timer system, which must be calibrated for the mist to spray at frequent intervals to reduce the chance of the plants' roots drying out. The water and nutrient solution is kept in a reservoir in which the mister assembly sits.

Aeroponic assemblies can be purchased for home use, but building them from scratch is a little tougher. Aeroponics depend heavily on a quality timer and misting system but do save money without the need to buy initial or replacement growing media. It's also difficult to start plants from seed in this type of setup because it's the most difficult to adapt for use with a growing medium. However, aeroponics can be a wonderful small system for growing kitchen herbs or greens, or use in a classroom. Many preassembled aeroponics gardens have a built-in grow lamp to supplement a lack of natural light.

The last of the six basic hydroponic designs is the **drip system**, which can be split into two categories- recovery and non-recovery. This technique requires a bit of engineering, but can be made from purchased parts, reclaimed or recycled components, or can be bought and assembled. The premise is simple- a reservoir holds the water and nutrient solution, a platform above this holds the plants in the growing medium, and a drip hose is suspended over the platform. A pump draws the solution up from the reservoir and drips it onto the plants, and the excess either drains back into the reservoir (recovery) or into a waste receptacle or drain (non-recovery).

There are pros and cons of each type of drip system, but both have strong upsides. When you use a recovery drip system, there is less liquid waste, but there is a greater need for regular upkeep; the nutrient solution may not be going down the drain, but it will need to be tested and refreshed more often. A non-recovery system can be filled and set on a timer, and since the liquid will not be recycled, it will not need to be rejuvenated. Drip systems can be as small as to fit on a countertop or be built to any size specified. Grow lights should be considered, as always, if there is a dearth of natural light.

With that overview of the types of hydroponics systems you can consider, let's take a look at the materials that are common to all or almost all of them:

- water

- water pump

- bubbler or air stone/pump

- reservoir

- platform and/or trays

- growing baskets and/or pots

- timer(s)

- grow lamps

- pipes and/or hoses

- growing media

- nutrient solutions

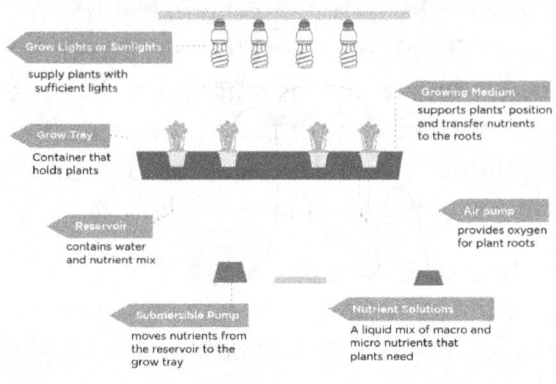

- the basic parts that are common to most hydroponics systems-

Many of the materials you'll need to build your system can be purchased at home improvement stores, hobby stores, or garden centers. If you are inclined to be a do-it-yourselfer, a lot of the enjoyment of a hydroponics garden is building it yourself. Many plans can be found on the internet for free or inexpensive download. If you're not into DIY, that's okay, too! There are a plethora of all-inclusive kits on the market for hydroponics systems, or you can ask or hire someone handy to build one to your specifications.

Growing Media

Growing media are the materials used to replace soil in hydroponics systems. There are a lot of growing media available to hydroponic gardeners, and each type is suitable for different applications. In order to make an informed decision about what kind of growing media fits you should choose, it is important to understand the options and their characteristics.

Expanded clay pellets- This growing medium is one of the most popular among hydroponics gardeners because it is versatile for many applications and in many styles of system. They are exactly what they sound like- clay that has been blown up into what looks like little terra cotta rocks, and they are used widely because they provide good drainage, dry quickly, can be sanitized and reused, and they create a stable structure for growing young plants.

The downside to these, as you can imagine, is their weight. Because expanded clay pellets are porous, they can be quite heavy when wet. Some gardeners choose to use them as a bottom layer in their growing baskets to provide stability and drainage while filling the rest of the basket with a lighter medium. Another downside is that the clay is a mined material, so there may be some issues with sustainability. You can always do your research into a company's mining and sourcing practices before purchasing their products if this is something that concerns you.

Coconut coir- Coir is a byproduct of the coconut industry, and is a popular growing medium due to its organic characteristics and its sustainability. It is also fairly inexpensive, although it does need to be replaced in between plantings; it can, however, be composted, given it is disease-free. Coir is the fiber that covers the outer husk of a coconut once it's removed from the woody shell. Coir is often sourced for things like doormats, but the smaller fibers are salvaged and used in things like potting soil and hydroponic growing media.

Coir is lightweight and compacted when shipped, and springs back to life when unsealed. It's often chosen as a growing medium for sowing and nurturing seeds and seedlings because it is naturally fungus resistant and holds moisture well. Coir can be used by itself, mixed with other media, or placed in biodegradable pots that can aid in creating the proper moisture balance. One downside to coir is that it can accumulate the salts out of nutrient solutions, but that can be alleviated by combining it with other media or placing it in pots instead of straight into a growing basket.

Perlite/vermiculite- These two minerals will be familiar to you if you've ever done any traditional gardening or if you keep potted houseplants. Perlite and vermiculite can be identified as the little white or gray balls you'll see in most potting soils; they can look like tiny spheres of Styrofoam. They are usually added to potting mixes because they are of a relatively stable pH, they take a long time to break down, and they create pockets of air that lead to less compaction and more stable happy root systems. In a hydroponics system, perlite and vermiculite can be used for this same purpose, sans the rest of the soil.

These media are wonderful alone, but because perlite is so light, it tends to wash away and is best mixed with vermiculite or other media. The two minerals together can be used, like expanded clay, as a stable drainage base, or they can be mixed in another medium like coir to balance out the weight and drainage. They are also mined from areas of volcanic rock, so if you are concerned about sustainability, you may want to choose something else or do research on suppliers.

Growstones- These are a medium that is becoming increasingly popular with hydroponic gardeners who are looking for long-term stability for mature plants. These pellets are made from recycled glass, are mid-weight, and provide a good water-to-air ratio for plant roots. Growstones are also relatively inexpensive, come from sustainable repurposing, and do not need to be mixed with another growing medium.

The one downfall of growstones is that they are not well-suited to starting seeds or for seedlings, because the tender roots are prone to wrapping themselves around the stones. However, most gardeners agree that growstones are an excellent choice for stability and nutrient balance in established, mature plants; they choose to use growstones in more 'permanent' systems.

Plugs- Opposite of growstones, plugs are best suited for starting seeds and nurturing seedlings. Plugs are small, bound cylinders or cubes of organic material, usually compost, designed specifically for sowing seeds or cradling plant starts. They can be used in almost every style of hydroponic system, they are fairly cheap, and they can be composted when spent. The only real downside to them is that you will have to transplant everything you begin in plugs to a more stable medium once the plants begin to grow and mature.

Oasis- Oasis is a non-organic compound that is used in hydroponics for starting seeds. If you're familiar with the floral industry or have ever gotten or made a flower arrangement, oasis is the dense, foam-like material used to hold the flower stems in place. It can be utilized in hydroponics because of its high water-retentive properties and its malleability. While it is single-use and non-compostable, it is inexpensive and readily available.

Rockwool- This is one of the oldest growing media known to hydroponic gardeners, and it's also one of the most difficult to explain. Rockwool is made up of very thin rock 'fibers' created by superheating stone and pulling it into thin strands like cotton candy or fiberglass. The result is a material that is made of stone, but which feels almost like fabric. Rockwool has been used in hydroponics for decades as a starter medium for seeds. Because it does not break down, it can be sterilized and reused for new plants.

Rockwool does need to be soaked before its first use, though, because the fibers can let off dust that may irritate the skin and lungs. People who work with rockwool over the long term often choose to wear gloves and dust masks when they handle the material. Rockwool is also difficult to dispose of because it doesn't degrade, and is not made in a sustainable manner. While it is a tried-and-true medium for hydroponic growing, it's up to you to measure the pros and cons and decide for yourself whether you want to use rockwool.

Those are the most common types of growing media used in hydroponic gardening, but there are gardeners out there who are getting creative and using other organic byproducts like rice hulls, wood shavings, and other cast-off materials. Some gardeners are going the cheapest route possible and using sand, pulverized brick, and gravel. If you want to use something other than a 'standard' growing medium, you should do some research. Some organic materials, like wood shavings, may invite insect pests. And repurposed building material like bricks or gravel may need frequent sterilization and replacement.

Now that you're familiar with some of the most common growing media for hydroponics, let's take a look at what goes into making a nutrient solution to feed your plants the minerals and compounds they would normally be getting from soil. This is the next step in understanding how all the components of a hydroponics system come together to provide your plants with everything they need to grow and thrive.

Nutrient Solutions

Once you've decided on what system you'll be using and the hardware you'll need to get started, and you've thought about the growing medium you think will best suit your needs, it's time to think about another piece of the puzzle, and that's the solution that will provide your plants with the nutrients they need to grow properly. While solutions will vary depending on what you want to grow, there are some fundamentals to every nutrient solution that should be aware of before you begin making formulas.

The first thing you should pay attention to when building a nutrient solution is your water quality. Water impurities can affect the growth of your plants, so it's best to have your water tested before beginning a hydroponics project. If you have softened water, it's possible that the salts used in treatment could have a negative effect on the pH or the mineral content of your water. If you have what's commonly known as hard water, it could contain lime or calcium that could also affect the overall nutrient content of your growing solution.

The vast majority of plants grow best at a pH of 5.5 to 7.0. While this usually applies to soil pH, in hydroponics, it applies to your water, and you should shoot for an optimal range of

5.8-6.3. pH is not used to determine fertility, but it does affect the availability and/or solubility of nutrients for your plants. It may be best for you to use distilled or desalinated water as a base for your nutrient solution, and take the guesswork out of the pH and eliminate the chances that trace minerals in your water will affect your plants. Picking up an inexpensive pH test kit and learning to use it will prove invaluable to you in your hydroponic gardening ventures.

In a previous section, we talked about the macro- and micronutrients your plants will need to grow strong and healthy. When you create a nutrient solution, you need to make sure that those nutrients are available to your plants in a proper balance so as not to cause deficiencies or toxicities. Most nutrient solution formulas measure the ingredient quantities in parts per million, or ppm. You can purchase most of your nutrients in a powder or liquid form to make your solutions. Formulas are easily found online or on the nutrient packaging.

You'll need to make sure that you've got your basic macronutrients- nitrogen, phosphorus, and potassium. These are usually found in their salt form, so you should familiarize yourself with the compound names ammonium nitrate, dihydrogen phosphate, and potassium nitrate. You will find that using the same brand for all your nutrients will make it easier to create a balanced solution, as the measurements will be consistent within the brand. Some brands even offer complete mixes or complete kits that can help you get started with your basic solutions.

When you create your first nutrient mix, be sure to follow the instructions very closely. Wear eye protection, closed toed-shoes, and gloves, as you'll be handling these nutrients in a highly-concentrated form. You should begin with the exact amount of water specified. You'll then want to dissolve your macronutrients- N, P, and K- in the measurements laid out by the nutrient manufacturer and your formula recipe. Once you've done that, you can move on to your micronutrients. Most commonly, you'll want to add manganese, boron, and iron.

Again, you should add these micronutrients only in the quantities specified in the formula. There can be too much of a good thing when it comes to feeding your plants. Once you've completed your mixture, you'll want to test the pH of your solution. If you're out of optimum range, you'll need to make adjustments. If your pH is too high, it means that you've made it too basic. The quick fix for this is to add white vinegar in small increments until you reach the desired pH.

The opposite action for a pH that's too low would be to add baking soda until you reach the optimal range.

One last consideration on the nature of your growing solution is to check its conductivity. Electrical conductivity in a solution that contains a lot of soluble salts can be a concern because too much conductivity will negatively affect the ability of your plants to take up the nutrient solution. A simple check with a conductivity meter or probe will give you your EC in a matter of minutes. The optimal range on this is 0.8-3.0, but you should ideally aim for an EC between 1.5-2.5. If your electrical conductivity is too high, you should add water in small increments and retest until you reach the proper number.

It may seem like there's a lot of work that goes into making a nutrient solution, but in all gardening, there is always a little bit of chemistry. As long as you can read instructions and follow them accordingly, you should be in good shape. Many of the components and the tools you will need are available at reasonable costs, and you can reach out on gardening forums or call your local Farm Bureau or Cooperative Extension for help should you need it. Many of the suppliers of hydroponic nutrients also have customer service lines to answer questions about using their products properly. The more you practice reading the formulas, making the solutions, and testing the pH and EC, the more proficient you will become.

Here's a checklist of the supplies you will need to begin creating nutrient solutions:

- distilled or desalinated water

- macronutrient (N, P, K) liquids or powders

- micronutrient liquids or powders

- a pH testing kit

- an electrical conductivity meter or probe

- vinegar and baking soda

- clean receptacles for mixing and storing nutrient solution

Now that you've got a good understanding of what goes into nutrient solutions and testing them for effectiveness, you're well on your way to planting and maintaining your system.

Chapter 2 Recap

This chapter was packed with information, so here's a recap of the main points:

- hydroponics systems are all based on the same principle

- there are 6 basic types of hydroponics systems upon which all designs are based
	- ebb and flow
	- wick
	- water culture
	- nutrient film technique
	- aeroponic
	- drip system
- the growing media you choose will be based on need and personal preference
	- each type of growing media has pros and cons
	- growing media can be mixed for best results
- nutrient solutions are all based on macro- and micronutrients
	- formulas can be adjusted according to plant needs
	- practice makes perfect when mixing solutions

Now that you've taken in all this information about the components of a hydroponics system, let's get down to business. In the next chapter, we'll go over setting your gardening goals, choosing your plants and seeds, and getting up, running, and planted.

Chapter 3: The Constant Gardener

As we've covered through the first two chapters, there are several components that make up a successful hydroponics system; the hardware and setup, the growing medium, the nutrient solution, and of course, the plants. That leads us to our next big question: What are you going to grow? What you would like to use your hydroponics system for will help you determine what type of system will best suit your needs, your budget, and your space.

Setting Your Garden Goals

The beauty of a hydroponic garden is that you can grow just about anything if you are willing to build a system to support it. Another upside to hydroponics is that since most systems can be built indoors, you don't have to worry about traditional growing seasons- if you want tomatoes in the dead of winter, you can have tomatoes in the dead of winter. Those two very positive characteristics aside, there are some considerations that you need to make when you're creating a hydroponics system to ensure that you're building the right garden for your needs. Here are some questions (with *sample answers*) that you can ask yourself to get started and set your gardening goals:

- What do I want to grow?

    - *vegetables and herbs*

- What will I do with what I grow?

    - *cook for my family and friends, donate to the food pantry*

- How much time do I have to devote to my garden?

    - *I'll be able to put effort into my garden after work every day and perform maintenance on the weekends*

- How much space do I have to build my garden?

    - *I've got an empty wall in the family room where I could house my system*

- Do I want to sow seeds, buy seedlings or starts, or begin with mature plants?

    - *I'll be doing sowing seeds for some varieties and purchasing seedlings for others*

- What is my starting gardening budget?

    - *I've saved up $500 to begin my set-up and acquire seeds and seedlings*

- What supplies will I need, and how can I get them affordably?

- I can build an ebb and flow system myself from repurposed parts, and I will purchase my growing media and nutrient ingredients from an online supplier. I'll get my distilled water, seeds, and seedling from local retailers. I can shop around for the best value for all of my purchases. I'll also need to research cost-effective grow lamps.

- How will I keep track of my gardening endeavors?

- I'll use an online garden tracker to record my purchases, planting and feeding dates, nutrient solution formulas, and other details so I can see what's working and what's not.

- Will I want to expand my system in the future?

- Yes, I think I'll start with a small table until I get the hang of things and then build a second one later.

When you sit down and answer these questions, you'll find yourself able to craft a plan to get yourself off to the right start. One of the biggest problems that plague new gardeners (or hobbyists of any sort) is getting in over their heads with a project that is too big or too complicated for their skill level. There is never any shame in starting small and building your way up to a bigger or more involved growing system.

Sourcing Seeds and Plants

It stands to reason that if you are going to put time, energy, and resources into building a hydroponics system, learning about choosing the best growing media, and investing in quality nutrients, you would also want to start your garden with the right seeds and plants for your gardening goals. If you've had or have a traditional garden, you know that low-quality seeds can result in poor plant growth and the frustration of lack of production. When you purchase seeds, there are some distinctions you should know about that will help you choose the ones that fit best with your goals and budget.

You should familiarize yourself with the following terms:

Heirloom- Heirloom plant varieties are exactly what they sound like. These plants are carrying the same genetic material as the generation before, and before that. Heirloom varieties allow botanists and gardeners to know precisely what they are getting when they plant a certain species, from the general size of the plant to the amount of fruit it should produce. These varieties are great because there is less guesswork as to how the plant should grow and behave, but they are not without flaws.

Heirloom varieties tend to be more susceptible to pests and diseases because of their lack of genetic diversity. In a closed hydroponics system, this is less of a concern because you will likely have your plants indoors. Heirloom varieties are also referred to as "open-pollinated" varieties. They will be able to self-pollinate or will be pollinated by plants of the same variety. The seeds of these varieties, when saved and planted, will produce plants of the same genetic characteristics.

Hybrid- Hybrid plant varieties are created when botanists and horticulturists choose the best traits of two or more varieties and combine them, either through breeding or grafting. The resulting varieties are usually hardier and more pest and disease resistant. Many traditional gardeners choose hybrid seeds for these reasons, especially in warm or wet climates where soil pathogens tend to thrive. If your hydroponics system is inside, and because there is no soil, you will not need to worry about these pathogens. You can choose hybrids because there's a neat variety you want to try, or you want to make sure the plant you're growing is hardy.

However, hybrid plants can be tricky to reproduce, because the seeds are often sterile. Even if you do manage to sprout new seedlings from saved hybrid seeds, there is no guarantee that the offspring will exhibit the same characteristics of the plant you saved the seeds from. Hybrids do not carry genetic materials in the same way as heirloom varieties.

GMO/non-GMO- If you are planting food crops, you'll want to make sure you understand the history and concept of GMO and non-GMO varieties. GMO stands for "genetically modified organism," and understandably, this can make people a little nervous. What this means, at its core, is that the plant material has been altered at the genetic level to exhibit a certain trait or traits. This is similar to hybridization but is done by changing the plant's DNA rather than through breeding or grafting.

Early GMO varieties were created to allow corn to be grown in more arid conditions to fight hunger in developing countries, and the first commercially available GMO food in the United States was the FlavrSavr tomato, which had its ripening enzymes inhibited to allow for longer shelf life. These days, GMOs tend to have a negative connotation, but the problems tend to arise more from the pesticides and herbicides needed to protect large GMO crops from the diseases and pests they tend to fall prey to. In a closed hydroponics environment, these issues are usually not present, and so you can choose GMO or non-GMO seeds based on your personal preference and comfort level.

The same discernment you use to choose seeds should extend to purchasing live seedlings or mature plants. Inspect all live plant material for signs of disease or infestation before bringing them into your home or garden. While you can often buy what some garden centers designate as 'distressed plants' for a steep discount, be sure that they are just in need of some TLC with water, food, and light, and do not have a pathogen or pest problem.

When you choose seeds, seedlings, or plants, there should be information on your seed packets or the plant tag denoting requirements and care instructions. This will indicate how much sun (full, partial, or shade) the plant needs, if the plants are light or heavy feeders (will affect your nutrient formula or frequency of feeding), and if the plants are suitable for container planting. If you are planning on transplanting to a traditional garden, then you can start about just any plant hydroponically. If you will not be doing so, you should get varieties that will do well in hydroponic containers even into maturity. Conversely, you can choose containers that fit the needs of the plants you want to grow, such as increased depth for plants with long roots or increased width for plants that typically spread out as they mature.

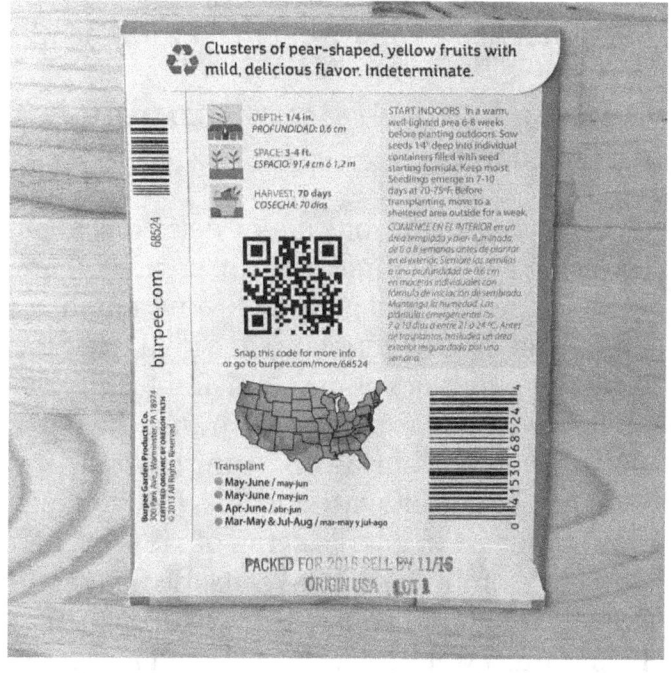

- seed packets have a wealth of pertinent information for plant growth-

One last note about purchasing seeds, and that's to be realistic. If you are going to grow a certain variety, how many are you going to need? Don't buy seeds in bulk if you are going to plant a few here and there- commercially available seeds only generally have a shelf life of two to three years before their viability will no longer be guaranteed. You can also plan to stagger your plantings (more on that in a bit) to make sure you've always got enough of your favorite plants growing without being overwhelmed. It's nice to have fresh tomatoes when you want them, but not nice to have so many you don't know what to do with them!

Getting Your System Running and Planted

Once you've made your major decisions about choosing a hydroponics set-up, created your gardening plan, and gotten all your supplies, it really is time to get planted! You'll want to begin by building your system and testing out all the moving parts before planting. If your system won't be sitting on an impervious surface like tile or synthetic flooring, you may want to invest in a large scrap of vinyl or linoleum, or an inexpensive plastic floor tray on which to build your set-up- hydroponics is by nature, a wet project, even when it's leak-free. There's no sense in running the risk of ruining a carpet or hardwood floor if there's ever a leak or spill.

Once you've laid your floor protection down and begin put together your rig, you'll want to test your air and water pumps. By making sure all your equipment works before assembly, you'll save yourself the frustration of finding out the water pump doesn't pump water before it becomes an issue. Be sure to save the user guides and warranty information for any equipment you've purchased new, just in case you have any trouble.

When you are building your system, be sure to follow the instructions, no matter if you are working from a kit or from raw materials with a plan or blueprint. You should test all the components that are supposed to be watertight to ensure that there are no leaks and make sure to use plumbers' tape or joint compound on any joints where water will be running through or sitting. If you are also installing grow lights, you should get those in place, too. Make sure you've hung or placed them at the height recommended by the manufacturer. As a general rule, you do not want your grow lights to be any closer than 12" from the tops of your plants to avoid scorching the leaves, but most light systems will come with specific recommendations.

Once you've got everything assembled, run some water through the system for a few days. Observe how the unit is functioning, and use this time to familiarize yourself with the controls on your pumps, bubblers or misters, and timers. Look for any areas of potential leaks, drips, or mechanical malfunction, and correct them before you do any planting. By running your entire system for observation and troubleshooting, you'll be comfortable with every component and their functions, which will aid you greatly should any problems arise in the future.

With your hydroponics system up and running, it's finally time to turn it into a garden. You should have your pots or baskets, trays, and growing media ready to go. If your growing media requires pre-soaking, you should be sure to do that in the same distilled or desalinated water that you'll be using in your system. You don't want to sabotage yourself before you start by introducing any elements that might create a nutrient imbalance. You should also drain any water you've been running through your system in anticipation of replacing it with your water and nutrient solution.

Finding an area big enough to work, lay out all your potting supplies. You should have your seeds or seedlings, growing media, pot or baskets, and trays all at the ready. Following a formula you've chosen, carefully mix your first nutrient solution and begin running it through the system. When it's been through the entire system once or twice (depending on your set-up), you can power the system down to rest while you plant.

When you plant seeds or seedlings, under any circumstances, you should follow the care recommendations for spacing to allow your plants the best chance to grow and thrive. This can be modified if you intend to transplant. In traditional soil-based gardening, many gardeners will begin their seeds in seedling cells or flats and then move them outdoors to harden off, then eventually into the garden bed to mature. It is then that they set them at the recommended distance.

Hydroponically-grown seedlings should receive the same care. If you find seedling flats that work with your system, that's wonderful! But you can just use small separate containers as well, to get your seeds off to a good start, if you intend to transplant them into a larger container later to mature. Following both the recommendations of your seed or seedling labels and the instructions on your growing media, you should be able to determine how best to plant your seeds to give them a successful start.

Armed with all your planting knowledge, it's time to get those seeds sown. Be sure to keep the seed planting depth recommendations and the proper use of your growing media in mind. Settle your seeds into their new homes and set them aside. When you've got everything down, carefully arrange the pots or baskets into their trays or holes (based on your system) and turn the system on. You want to make sure nothing has started leaking since your initial tests, and see that the water and nutrient solution is reaching all the plants the way it should.

If you have purchased seedlings that were grown in potting mix or transplanting mature plants from outdoors or from a container, you will need to clean the roots before settling them into a new hydroponic home. You'll want to gently remove the plants from their current container and remove as much loose soil as you can with your hands. You can wash the roots off in a bucket of water and then give them a gentle rinse under a running faucet or sink sprayer to get them as clean as possible. The idea is to not introduce any foreign matter into your hydroponics system, or you may run the risk of clogs or contamination. Once the roots are as clean as you can get them, you can feel confident knowing that you did your best to keep your system pure.

Timing and Timers

Once you've got your seeds and seedlings planted, and you're sure the system is leak-proof and running efficiently, you'll want to set your equipment up to run when it's appropriate for maximum effectiveness. Most pumps and grow lights have built-in timers, so you can 'set it and forget it.' The optimum situation would be to have your entire system run at the same time, so if the settings on all your devices do not allow for this, you may want to get a power strip with a timer or an outlet timer which you can plug the devices into.

Most young plants need at least eight hours of light to grow at the correct rate, and you can usually go up to twelve hours without harming the seedlings. If you also want to feed and water the plants for twelve hours, then your timing is simple- you can set up the system to go on in the morning and turn off at night, from 7 a.m. to 7 p.m. for example. You could also run in four six-hour increments, if you like, six hours on and six hours off, twice a day. As long as your plants are getting adequate light and nutrients every day, you can choose a schedule that works for you. Remember that tender young plants need time to rest and work on cellular regeneration just as young humans do.

If you have no faith in timers, you can, of course, turn your system on and off manually every day. This is a great way to remind yourself to observe your plants daily for growth and monitor for any issues. One note about manually operating your system; if you are going to be traveling or away from your garden for a few days, you should set it up with fresh nutrient solution and set a timer so that your garden if self-sufficient or low-maintenance for a house-sitter while you're gone. You can always turn off the timer when you get back.

Planning for Succession Growth

One of the best things about having an indoor hydroponics system is that you don't have to be held to traditional growing seasons. That means you can grow what you like, when you like, as long as you can provide the proper conditions for your plants. Being able to have fresh food and herbs to eat and fresh flowers to enjoy year-round is a huge reason that many people choose to begin hydroponic gardening. In this section, we'll be focusing mostly on food plants as an example, but the principles will apply to anything you may want to grow in succession.

Most food-producing plants are annuals, which means that they will live and bear fruit for one growing season, and then the plant will consider its job to be done, and it will begin to die off. Remember that fruit, which protects the seeds, is the final offspring of the plant. It has matured (grown and flowered), been fertilized (been successfully pollinated), and produced children (seeds protected by fruit). This is an annual plant's purpose in life, and it will die a natural death once it has fulfilled it.

Because of this, many traditional gardens are planted in three 'shifts'- plants that thrive and flower in the cool weather of spring, warm weather crops that are harvested at the peak of summer or shortly thereafter, and another set of cool-weather crops that will be ready to go right before or around the first frost. Interplanted with these crops are the long-maturing varieties, like pumpkins or gourds, which can be planted early but take the entire growing season to mature and produce viable fruit.

When you are growing food hydroponically indoors, you can adjust what would typically be an annual plant's growing season, but you cannot adjust the nature of their lifecycle. In other words, you can plant green beans in the dead of winter, but you cannot make that bean plant live longer than one growing season. Once it has produced its fruit, it will feel that its job is done. One way to make sure that your food-producing plants will last as long as possible is to harvest regularly, so the plant isn't overburdened with putting energy into feeding overripe fruit. You'll also avoid a lot of mess and waste by harvesting properly and promptly, as well.

By reading the information on your seeds and plant tags, you will know how long it takes for your plants to mature and reach harvest. You can use this data to make a succession plan for having fresh food year-round without having an overabundance that may go to waste. You should plan to plant varieties that have a mix of harvest times when you start out because that will give you the opportunity to stagger your plantings of new crops. Start by making a list of the things you've decided to sow and mark down the 'days to harvest' next to each variety. Remember, you don't have to grow everything all at once. Choose what you'd like to get started and be creative. Your garden plan should have included your purpose, and you can grow accordingly. If you wanted fresh food for your family, but you have teenagers that love pizza, grow pizza items- tomatoes, garlic and onions, and basil and oregano. If you want to make smoothies, you can grow kale, strawberries, and carrots. The only limit is your imagination.

Once you've established your garden, you are now in control of your succession plan. You will be able to harvest the fruits, vegetables, and herbs from your system as needed and at will. When a plant is spent, you can remove it and plant a new seedling in its place. Remember that if you are utilizing a reusable growing medium, that it will need to be thoroughly cleaned and sanitized before it is planted again. You don't want to run the risk of cross-contaminating your growing media or your nutrient solution.

The beauty of succession planting in a hydroponics system is the freedom that you have to change out plant varieties and try new things. Be sure to keep track of your planting dates in your garden journal or log, so that you can track your plants' growth and how they are responding to your nutrient solution. You'll also want to observe how your young seedlings are responding to your light settings, to make sure they are not getting too much or too little. When you make and notate your daily observations, you can make corrections as needed before any concerns get out of hand.

If you see signs that your seedlings are yellowing or experiencing leaf drop, your solution may be too strong or too weak. You should test the pH and EC, and adjust based on your findings. If you see 'leggy' seedlings, they are likely trying to find better access to a light source. Consider moving your grow lamps closer to the seedlings or raising the intensity setting on them. If your seedlings are curling or brown, they may be getting scalded by your lights, in which case you can move them farther away or turn them down a setting or two. Being proactive and observant can make all the difference in a plant's success or failure.

We've touched upon keeping a journal, and this is a valuable tool. Keeping a journal means you'll know exactly what days you've done all of your gardening tasks- refreshing your nutrient solution, planting and harvesting, adjusting lights, and other regular activities. By tracking all your activities, you will be able to craft the big picture story of your garden, know if you are meeting your gardening goals, and be able to make amendments as needed to be successful in your hydroponics venture.

Chapter 3 Recap

We covered a lot of ground in this chapter about getting your system ready to go and getting planted. Here are the main points you should follow:

- set your gardening goals

- lay out all supplies for building and/or planting before you begin

- test your system for functionality, leaks, and other issues before planting

- familiarize yourself with your hardware

- choose quality seeds and plants, and read labels for pertinent information

- plant your seeds/seedlings according to recommended depth AND growing media instructions

- keep a garden log or journal either by hand, via a gardening application or software, or spreadsheet

- plan for succession planting to ensure constant growth/plants for harvest

- observe and troubleshoot regularly and document your concerns and successes

Congratulations on getting your system up, running, and planted! You've done a lot of hard work so far- being a researcher, builder, scientist, writer, and gardener. To protect everything you've accomplished, you're going to want to learn to be a maintenance person, too. Because no system is perfect and because you don't want to let little problems become big ones, we'll spend the next chapter going over the best ways to keep your hydroponics system humming and growing in top condition. Regular maintenance and the ability to find and fix issues will help you make sure that your system will run smoothly for years to come.

Chapter 4: Maintaining Your System

A hydroponic garden is so much more than just a garden; it's a machine and a system. Like all mechanical and electrical systems, it will require maintenance and attention to stay up and running in an optimal fashion. In this chapter, we'll address regular maintenance routines for both your plants and your hydroponics, some basic troubleshooting, and tips on keeping your plants and your system happy and humming.

Happy Plants, Happy Harvest

The main goal of all gardening endeavors is to grow plants that are healthy, productive, and happy. What are happy plants? Happy plants are those that get the correct amount of the things they need- water, nutrients, light, and air. When you designed and planted your system, you did so with the best interests in mind for growing healthy, happy plants. Once you've got your system running, what's the best way to keep those plants on track?

You should regularly inspect your plants, daily if you can (and what gardener doesn't love looking over their seedlings with that parental feeling of pride?). You want to look for signs of nutrient deficiency or toxicity, check growth to see if your light source needs to be adjusted, and look for any signs of disease or pests. While insect pests and infection are unlikely in hydroponic gardens, they aren't impossible. Daily observation will help you find any issues and alleviate them before your entire garden becomes affected.

You need to remember that your nutrient solution is not a one-and-done task. Its job is to nourish your plants, and they will be taking up the components that they need to grow strong and healthy. Test your solution often for pH and EC. This will not tell you the content of elements left in the solution, but it will tell you how the plants are affecting the nutrient balance. If you see very little change in pH and EC from your initial readings, but see the total volume of liquid in the system going down steadily, then your plants are feeding evenly and efficiently. You should plan on changing or replenishing your nutrient solution every two to three weeks, and if all is going well, you can likely stick with that same nutrient formula and just refill the supply. If you see a drastic change in pH and conductivity, your plants may be exhibiting uneven feeding.

- healthy salad greens grow in a well-balanced hydroponic system-

If you find that your plants are showing signs that something may be off with your nutrient solution, you should test the pH and EC and adjust your solution accordingly. You may need to drain the system and replace the entire supply of solution if the balance is too far off to correct. As a reminder, the most prevalent symptoms of nutrient imbalance are leaves yellowing, browning, curling, or dropping excessively.

Another thing you should check on a regular basis is your plant roots. If you are not planning on moving your seedlings to a traditional garden, you may find that they will become root-bound in their original containers. If this is the case, you should gently free them and transplant them into large baskets or pots. Root-bound plants have a difficult time taking in nutrients through and may begin to wither or get root rot, which normally would not be an issue in a hydroponics garden, given that the system should be timed to give the roots a while to dry out between feeding cycles. Root-bound plants can retain too much moisture in a hydroponics system and invite decomposition.

One common issue in hydroponics is the growth of algae on surfaces and plant roots. The algae growth can clog up your system and upset your nutrient balance, so you should be sure to keep wet surfaces clean with regular wipe-downs and don't let algae build up. Unfortunately, algae concerns are not a question of if, but when, so make sure you don't allow it to become a big problem when it does occur. Just be vigilant, wipe hard surfaces, and don't create an environment in which algae can thrive. You can do this by making sure your system doesn't sit stagnant, that you have good airflow, and that you refresh your nutrient solution when appropriate.

If during your regular inspections, you discover that you've got a pest or disease problem, you'll want to immediately remove the affected plant(s) from the system and assess if they can be treated or need to be disposed of. The internet has a wealth of information about common garden pests and plant pathogens to help you identify your invaders, and you can familiarize yourself with the issues most common to the varieties you are growing. Your local Cooperative Extension or Farm Bureau likely also offers identification services and recommendations for remediation.

If your plants can be salvaged, take the appropriate steps to do so. You should preferably try using as few chemical products as possible, especially because the plants are likely inside your home. Try organic remedies like insecticidal soaps or natural fungicides. For some insect pests, like aphids, a quick, hard rinse of water will send them down the drain for good. Remember that if you should need to dispose of any plant matter that has been subject to pathogens, you should put it in the trash rather than a compost heap to avoid the risk of spreading the disease to any outdoor plants.

No one wants to have to make the decision that a plant is no longer viable, but you should cut your losses and avoid infecting or infesting the rest of your plant population. If you do experience a pest issue or a pathogen, make note of what plants were affected, when, and what action you took. You should also try to determine what caused your issue.

Perhaps it was something as simple as insects making their way in through a hole in a window screen; that's an easy diagnosis and an easy fix.

Maybe your seeds or seedlings were carrying a pathogen that was unseen or unable to be seen at the time of purchase. You make a note to either find a different supplier or be more vigilant when using that supplier again. Try to give vendors the benefit of the doubt, because no one is perfect. Plant and seed purveyors are not actively trying to sell you anything defective, as that will also affect their reputation and their bottom line. If you suspect that faulty seeds or seedlings may have been the cause of your plant's problems, offer the company or supplier that information diplomatically. They may have other customers experiencing the same issues, and it will help them pinpoint what went wrong at the source.

If you experience a pest or pathogen issue, learn from it! It doesn't necessarily mean you've done anything wrong- gardening is an imperfect hobby. You can learn and grow your knowledge and expertise while you build and grow your garden. As long as you remember the basics- water, air, light, nutrients- you will have many more successes than failures.

Maintaining Your Machines

The other important things to regularly observe and maintain are the mechanical components of your hydroponics system. You want to make sure that everything is functioning in peak condition and running efficiently. You likely have one or more motorized components, like a water pump and a bubbler. You've also likely got some grow lights, and they may need new bulbs occasionally depending on the style you're using.

The one thing to remember when you are combining running motors and water is that electricity and water do not usually make good partners. You'll want to regularly check to ensure that the housings on your pumps are not cracked and that their cords are not split or frayed in any way. If you like, you can put cord protectors on them, which are available from most home improvement or contractor supply stores. Be sure to also check the cords on your grow lamps for any damage. If you find anything concerning, be sure to remedy the situation immediately or take the damaged machinery out of your system.

When you were going through your initial set-up, it was advised that you run the system for a while to familiarize yourself with the components and how they normally sound and behave. One wonderful human attribute is that we tend to know how all the things in our homes sound, and can identify almost immediately when something seems 'off,' like the refrigerator making an odd clunk or the furnace making a strange noise. The same should be true for your hydroponics system. If you've become accustomed to the noise it *should* be making; you'll know right away when it's not.

You should also have saved any owner manuals or user guides for your mechanical and electrical equipment. If something goes haywire, these manuals usually include troubleshooting guides and manufacturers' contact information to help you make a diagnosis at home or decide to find a qualified repair person or facility. The information in user guides can prove invaluable because it is specific to your make and model of equipment. Many companies also provide websites with additional information that will also be of assistance to you, and live text-based chat is also becoming a popular customer service feature.

If you have no experience with repairing your own machines, please do not run the risk of injury or property damage. You can find someone with the expertise and ask them to teach you how to do it yourself, or find a trusted service provider to do the work for you. If all else fails, you can upgrade your equipment with a new model. The most important thing is to have equipment in good working order all the time.

You should also check your equipment regularly for drips, leaks, and clogs in the plumbing. You don't want to lose valuable nutrient solution to a mess on the floor, which can also cause property damage and frustration. Clogs can back up the system and burn out your pumps, which is another big concern. If possible, you should flush your system a few times a year by removing the plants for a cycle and letting the system run without roots in the way. This is so you can observe your system in much the same way you did when you first set it up. Take care of any concerns promptly, and get your plants nestled back into their homes.

Should you have a power outage, there are a few scenarios that can play out. In the first scenario, your power went off during your systems 'on' cycle, and there is some nutrient solution laying in the plumbing or the plant pool. This isn't harmful if the power outage isn't very lengthy. When the power is restored, the system time will either reset itself, or you should manually reset it. It may be best, depending on how long the outage was, to run the pump long enough to drain the system and restart it the next day.

If your power went out during the 'off' cycle, you'll have a similar problem, but in reverse. Just as you don't want your plants sitting in their solution for too long, you also don't want them to go too long without, either. If the outage is short, it shouldn't be much of an issue. You can just give the roots a nutrient bath when the power comes back on and reset your timers for the next day. After either situation, you'll just want to double-check your equipment to make sure there was no damage from an electrical surge.

In the case of a lengthy power outage, you may need to take additional steps to ensure that your plants will not be without their requirements for too long. If you have a generator, you may want to make sure your system is hooked into it, but again, check your equipment for damage from surges. If you don't have a generator or access to a generator, you will want to find a way to manually care for your plants during your extended outage. Another option might be a battery-operated pump that you can use in place of your corded pump. Because light may be an issue, you will want to open up blinds and shades to allow your plants to get as much natural light as possible, and come to terms with the possibility that your plants may be a little stunted.

Keep It Clean

Good housekeeping means less chance of contamination and clogging. Of course, dusting is far from everyone's favorite chore, but keeping your system clean and free of dust and contaminants is good practice. This is especially true if you have house pets whose fur could clog up the air filters on your machinery or get into your plumbing components and cause hair clogs. To avoid this, you should wipe down your system a few times a month and run a vacuum on the air filters of your pump(s).

Because your plants are indoors, they may also be gathering a little dust on the leaves. Wipe them gently with a soft cloth to make sure your plants can breathe easy- they respirate through tiny stomata on the underside of their leaves. If you don't want to breathe in a lot of dust, neither do they. Just performing the simple chore of keeping the leaves dust-free will help them better perform their basic biological functions.

Chapter 4 Recap

Congratulations on making it through the maintenance chapter! There was a lot of information here, so let's recap before we move on to our last chapter.

- plants should be checked regularly for the following:

    - signs of nutrient deficiency or toxicity

    - signs of needing more or less light

    - being root-bound/needing transplant

    - signs of pests or pathogens

- nutrient solutions should be tested and replenished often

- mechanical and electric components should be monitored for damage

    - refer to manuals or guides for troubleshooting assistance

    - be attentive to electric and water components to avoid problems

- all damage to plants or parts should be mitigated as soon as possible

- have a contingency plan for power outages

- keep your garden and your garden area clean with basic housekeeping practices

Now it's time to head to our final chapter, where we'll talk about ways to expand your hydroponics system and give you some unique ideas for making your garden healthier and more productive by delving into a little bit more of the science behind plant growth and nutrient solutions. We'll also talk about some of the things you can do to enjoy and preserve your harvest, be it food or flowers. Let's go!

Chapter 5: Next-Level Hydroponics

You've made it to the last chapter! By now, you've got your garden started, or you have a really good idea of how to do so. In our final sections, we're going to talk about ways to step up your gardening game, either by expanding your system, getting into more advanced nutrient solutions, or finding a value-added hobby to help you get more enjoyment out of your garden. It's time to take your garden to the next level!

Growing Up and Growing Out

There's a little bit of a problem with gardening, sometimes, and that's that it can become a bit of an obsession. Success breeds success, so if you find that it's time to expand your hydroponics, what are your options? Can you add to your existing system? Do you want to try something new with a different type of system or new growing media? Just like when you set your initial gardening goals, it's time to set your expansion goals.

If you have space, you can always just add a new system alongside the one that you already have. But that's not always ideal, and before you know it, you may have taken up too much valuable space in your home. How about growing up? Vertical expansion is always an option. Consider adding another reservoir shelf and a stronger pump to an existing ebb and flow system. A little simple plumbing reconfiguration can help you add new pipes to other types of systems, as well.

Be creative when you are adding to a system. Draw out your existing pipes and pools and play around with it on paper- it's always easier to erase pencil lines than erase plumbing mistakes. Think about the space you have and then think about the ways you can maximize it. Grow up, grow out into a room, grow along the length of a wall? What's going to make the most sense for your space and your needs? Take physical measurements and see exactly what you have to work with; it will give you a much better sense of the space when you're sketching your ideas.

No matter what you choose to do, make sure that when you are adding to an existing system, you take the same care in design and set-up as you did in the beginning. Check for leaks and drips, and always test your mechanical components for functionality before adding plants to the system. You'll want to be especially attentive to any new pipe joints and reservoir trays. If you've expanded vertically, be sure that any new pumps are strong enough to circulate your nutrient solution through the entire system, and check that your air bubblers can also handle any added capacity, as well.

The key to expanding or adding a system is to know what your personal gardening capabilities are, too. Don't expand for the sake of expanding, because you don't want to burn yourself out and be stuck with a large system you cannot maintain- that just adds stress and takes all the fun out of it. Grow things that you will use and enjoy, and you'll keep the fun in your endeavors.

Tweaking Your Nutrients for Maximum Growth

After you've had your hydroponics up and running for a while, you may be noticing a pattern in which plants are growing better than others. This may be because some plants are heavier feeders than others, meaning they need more nutrients to maintain growth and production. If you are growing a lot of heavy feeders- things like tomatoes and peppers- you may need to adjust your nutrient solutions to give them a boost. Conversely, if you are growing light feeders- lettuces and greens- you may be giving them too much of a good thing, and you'll want to dial back on the amount of food you are giving them.

In this section, we'll go over how to adjust your solutions to better fit your plants' needs and how to add fertilizers to your system when it's appropriate. Because you are the one in control of making your solutions, you also have control of how strong they are and when it's time to modify those concentrations. First, let's look at some of the plant families and their more specific nutrient needs.

Heavy feeding plants are those in the nightshade family (tomatoes, peppers, eggplants, etc.), those in the brassica family (cabbages, broccoli, Brussel sprouts, etc.), and some root vegetables like beets. These plants take in a lot of nitrogen when they feed, and will deplete this element from your nutrient solution very quickly. If you are growing plants in these families, you will have to increase the nitrogen in your nutrient solution or refresh it often in order to fill the needs of these plants. Other heavy feeding plants include corn, cucumbers, and most melon and squash varieties. It takes a lot of nitrogen for these plants to develop their thick stems and vines.

If you are going to be growing light-feeding varieties, you can stick with a generic formula for your system, or even reduce the nitrogen if you see early signs of toxicity. The plants include most lettuces and herbs, as well as beans, peas, and other legumes. The beauty of legumes is that they also put nitrogen back into the system. In a traditional planting, this would be returned to the soil in the form of exudates from the roots and a unique symbiotic relationship with a soil microorganism known as rhizobia. These bacteria will not be present in your hydroponics system, but at least you will know that your leguminous plants will not be rapidly depleting your nutrient solution.

Most flowers and bulbs fit into the medium-feeding category. You can use a standard solution for these plants and adjust accordingly after you observe how the plants are behaving. One thing to remember, everything in your system should be in balance. If you plant some tomatoes, also plant some beans. Try to find a way to alleviate the rapid depletion of nutrients in your solution if you are planning on planting a variety of species. If you are planting a monoculture, then you can simply use the solution that works best for that variety.

PERIODIC TABLE OF PLANT NUTRIENTS

7	15	19	12	16	20
N Nitrogen	**P** Phosphorus	**K** Potassium	**Mg** Magnesium	**S** Sulfur	**Ca** Calcium
Primary Macronutrients			Secondary Macronutrients		
5	17				
B Boron	**Cl** Chlorine				
25	26	28	29	30	42
Mn Manganese	**Fe** Iron	**Ni** Nickel	**Cu** Copper	**Zn** Zinc	**Mo** Molybdenum
Micronutrients					

- an easy reference chart of plant nutrients-

The above chart is an easy reference for the elements that plants need to grow and thrive. When you want to grow a specific variety, and you're not sure if it's a light or heavy feeder, you should look it up before planting, so you know where it fits into your garden. Remember that you are not only creating a space for plants to grow, but you are also creating an ecosystem that needs to stay balanced to be healthy and productive. As with all ecosystems, balance and diversity will create an ideal atmosphere for growing your plants.

When you start researching formulas for nutrient solutions, you'll find a whole world of studies, data, formulas, and other scientific information that may make your head spin unless you're a chemist. Don't freak out! The only thing you have to do is find a formula that's appropriate for the results you want and take your time. Read carefully, and be aware that it's all based on ratios. Most formulas will give you a measurement for parts per million (ppm) based on using a certain quantity of water, most commonly five or ten gallons. Do yourself a favor and get good quality measuring equipment, and remember not to rush, and your formulas will come out just fine!

One thing to note is a frequently asked question in hydroponics: can I use regular fertilizers to make my hydroponics solution? The answer is that it is highly discouraged. You should only use products indicated for use in hydroponics systems. These are formulated to have the correct concentration of macro- and micro-nutrients for soil-free gardening. Most traditional fertilizers only contain N, P, and K, and do not have the micronutrient content necessary, because these trace elements are normally found in most garden soils. Using NPK fertilizers not meant for water solubility can also lead to clumping, clogs, and messes in your system, which is obviously not desirable to have to deal with.

The last thing to remember about nutrient solutions for your hydroponics system is that you need to give them time to work. Don't rush to judgment on a formula if you don't see results overnight. While a wilted plant may show immediate signs of perking up when given water, remember that your plants need time for the nutrient solution to penetrate the cell walls inside your plant and take action. Give it a few days, unless you see immediate signs of distress and toxicity. Always keep in mind that less is better; you can always give a plant more nutrients, but it's very difficult to remedy an overfeeding.

Using Your Harvest Today and Tomorrow

Now that you've grown your plants, maybe even expanded your system, and decided to explore different nutrient solutions, you're probably looking at a pretty abundant harvest. What are you going to do with it all? Let's explore some ideas for expanding your garden hobby in our last section. It's always good to end on a fun, creative note!

If you are growing food in your system, chances are good that you'll be enjoying your harvest in its freshest form as it grows. The health benefits of having fresh, available produce cannot be denied, and the thought of making a salad that you harvested in your living room is appealing and wonderful. But if you have too much fresh produce, what can you do to make sure nothing goes to waste? You can consider canning, freezing, or drying your fruits and veggies so you can enjoy them later. Many people are employing these methods to keep their harvest safe and enjoyable for a later date.

When you take up canning, the biggest consideration is food safety. There are two types of canning, water bath canning, and pressure canning. The first is a more traditional method and can be used to preserve low-acid foods and pickled foods. Water bath canning involves preparing your foods, either raw or processed into jams, jellies, salsas, or pickles, and placing them into sterilized canning jars. The lids are then fitted, and the jars are lowered into a bath of boiling water which seals the lids and disallows air from getting in and spoiling the food. Pressure canning requires a specialized canning pot that is fitted with a lid and valve to seal the lids with vacuum pressure. This is ideal for all foods, but anything considered high-acid, like tomato sauces, soups, and stews containing meat, MUST be pressure canned to assure food safety.

You can purchase canning jars and lids almost anywhere, from grocery stores to hobby stores, and even at large department stores. The key thing to remember when canning is that the jars and lids must be absolutely clean and sterile when you begin. Jars can be boiled empty or placed in a 225* oven for ten minutes, and the single-use lids should be washed and rinsed in lukewarm water. Use a canning magnet and tongs to avoid touching the jars and lids after cleaning, and always wipe the rim of each jar with a soft clean cloth or paper towel before fitting the lid. Leave a little room, called headspace, when you are filling jars to allow for expansion. When you remove your jars from the canning bath or pressure canner, do not touch them for 24 hours, so that you do not disturb the fresh seal and so that the jars have time to rest and cool. You can place the jars on a tray and set the tray somewhere safe for a day while they cure. Always label your canned goods with the contents and canning date.

If you're not comfortable with taking up canning, freezing can also be a way to save your harvest for later. This is a great method for putting up fruits and vegetables that you might normally purchase frozen anyway, things like strawberries or green beans. If you are going to be freezing fruit, be mindful that the high moisture content may make them a little soggy as they thaw. You should freeze fruits whole if possible, in double freezer bags, labeled with content and freeze date. Freeze fruits as soon as possible after harvest to preserve as much of the nutrient content as possible.

When you freeze vegetables, you may want to process them first to save time when you want to eat them. Chopping and blanching is the best way to preserve the nutrients and create a usable product, especially for green vegetables, which tend to leach out their vitamin content when they are boiled. For example, if you wanted to freeze peas or green beans, you would shell the peas (unless, of course, you want them in the pod) or snap the beans into manageable pieces. Prepare a lightly-salted boiling water bath and a bowl of ice water.

Using a fine strainer or steamer basket, dunk the veggies into the water for no longer than it takes for them to turn bright, vibrant green, usually about a minute or two. If you don't have a strainer, you can use a slotted spoon. Remove the vegetables from the hot bath and put them immediately into the ice bath. This stops the cooking process in its tracks. Once the veggies are cool, drain them very well, patting dry if possible, and put them into a labeled freezer bag and into the freezer. When you want to eat them, you'll just need to take them out and prepare as you would any commercially-available frozen vegetable.

Some foods are perfect for dehydrating or drying, like herbs, peppers, and beans. Some stoves have a dehydration setting, and some people have countertop food dehydrators, but you don't need anything fancy to utilize some basic drying techniques. All you need is time, air, and a little patience, and you can have your very own herbs in the spice rack and plenty of beans for making soups and stews all winter long.

When you want to dry herbs, one of the best tried-and-true methods is to hang them upside down somewhere dry and relatively warm. You can hang a twine 'clothesline' for doing this, complete with little clothespegs if you like. Hang your herbs in small bunches, leaves down, stems up. You'll see progress every day as they become drier, and when you can give a leaf a good test crumble, it's time to take them down and crumble them up. The best storage for dried herbs is in airtight glass jars, which you can find at any hobby store.

If you want to dry vegetables, you can also use the hanging method, or you dry them on a tray lined with newsprint or paper towels. If you lay out things like peppers or beans, be sure to turn them every day and change the absorptive paper under them regularly. When they have no moisture left, you can store the peppers whole or grind them into a spice jar. Beans can be shelled and stored in airtight jars until you're ready to use them. As always, remember to label everything with the contents and the date you stored it.

Because food isn't everything, what should you do with any flowers you've grown that might not make it into a fresh bouquet or flower arrangement? You can also dry them, for use in potpourri and sachets, or press them for use in arts and crafts projects. Saving flowers for use at a later time is a great way to create beautiful, useful gifts and mementos, and can be a rewarding hobby in and of itself.

If you want to dry flowers, you can do this in the same manner in which you dried your herbs, hung upside down. For all you science buffs, drying the plants upside down allows the moisture to stay the longest in the most important part of the plant- the flower- so that it maintains its structure as it dries. Most flowers take a week to two weeks to dry out properly. After this, they can be lightly sprayed with a coat of shellac and used for dried flower arrangements, or the petals can be crushed and mixed with fragrant wood shavings or spices for potpourri. You can also sew the mix into fabric scraps for sachets to keep drawers and closets smelling fresh.

If you'd like to press flowers instead of drying them intact, you'll need some big heavy books, two pieces of wood and some sturdy rubber bands, or a flower press, but there's no need to purchase one if you've got the books or the lumber. For pressing flowers, you should remove as much of the stem as possible and place the bloom in some newspaper or tissue paper. Then place the wrapped flower into a book and close it, weighing it down with another book. Or you can place it in between the two pieces of wood and rubber-band them tightly together. Check on your flowers every few days, replacing the paper when necessary. When completely dry, the pressed flowers can be used in art projects, or placed into poured homemade soaps and candles. Pressed flowers should be stored in airtight containers to protect them until use.

You can surely think of a lot of other things to do with your harvest, but these are just a few ideas to get you started. You can also check with your local food pantry to see if they accept fresh produce, or bring flower arrangements to nursing facilities to brighten people's day. The possibilities really are endless. Once you have your hydroponics system running and productive, nothing is stopping you and your creativity. Have fun, and try new things, and you'll be amazed at what you can do.

Chapter 5 Recap

Welcome to the last recap of the book! It's been a great gardening journey with you, and we hope you learned a lot. Here are the main points from the last chapter for you to review.

- when expanding your system, ask yourself the following:

    - different type of system or build an extension?

    - what are my space considerations, and should I go up or out?

    - what's my personal capacity for gardening?

- find ways to expand your knowledge of nutrient solutions

    - have you planted a monoculture?

    - if not, have you planted a good balance of heavy, moderate, and light feeders?

    - consider what you can do to improve your solutions for stronger plants

- don't overfeed- more is not always better!

- plants need certain elements to thrive, and you are in control of your plants' needs

- unbalanced feeding will lead to plants' failure to thrive

- nutrient solution formulas require time and patience, not a science degree

- traditional fertilizers should not be used in a hydroponics system

- learning to preserve your harvest will lead to less waste

- can, freeze, or dehydrate food for later use

- dry or press flowers for use in arts and crafts projects

- consider donating any food or flowers you cannot use or preserve

Now that we've come to the end of the book, we sincerely hope that you've gained all the knowledge you need to get started on your hydroponic gardening adventure! Remember, the key is to be prepared and to be thoughtful. Sketch your designs and price out materials. Try to source used components to keep your initial costs down. Seek out the advice of other hydroponic growers if you need assistance. Don't be afraid to start small, and always follow manufacturers' recommendations on seeds, equipment, and hardware.

Mostly, we want you to enjoy your garden. It's a hobby that's proven to be good for your health, even before you eat a fresh veggie straight from your living room. Gardening keeps the mind sharp, can lower stress levels, and connects people with a process that delivers a tangible result. Gardening is good for everyone, and everyone can garden. Hydroponic systems can deliver that ability to the housebound, bring fresh food to the plates of your family, and grow flowers to bring you and your loved ones much joy. There are not too many hobbies better than that!

Now go out there and get growing! There are no limits to what you can achieve with a hydroponic garden when you put your mind to the task. Be a grower, a researcher, a chemist, a reader, and a writer; challenge yourself to try new things and grow new varieties. And when you need a little help, come right on back to these pages and refresh your memory. We can't wait to see you again. Happy gardening!

Conclusion

Thanks again for reading *Gardening Secrets*, your ultimate guide to getting started and succeeding in hydroponic gardening. We hope you found the timely information and helpful tips to be exactly what you were looking for! We covered a lot of ground, and all the knowledge here should be more than enough to get you on your way to a bountiful harvest.

One of the best things about taking up gardening is that there is always so much more that can be learned. Gardening techniques are constantly evolving as plant science discovers more about the way plants can take up nutrients and how those nutrients are used within the plant's cellular structure. For hydroponic gardeners, that means that the future holds even more advanced nutrient mixes that will allow for bigger, more beautiful plants.

On the flip side, gardening is also amazing because it is rooted in so much tradition. People have been growing and harvesting food since pre-history, and being part of that can keep you connected to the past and the people who gardened before you. Every time you start a hydroponic project, you are honoring the ancient engineers who built the hanging gardens, you're paying homage to the research of Dr. Gericke in the 20th century, and you're making your own garden history in the 21st. How will you approach your project to make it stand out? It's fun to think about the possibilities.

This book offered you a lot of science and engineering, and we hope you gained a lot of insight into plant structure, function, and nutrient requirements. If you're interested in learning more about botany or organic chemistry, there is a wealth of information to be had on the internet and by taking workshops and classes- you can check with your local library, Cooperative Extension, or Farm Bureau to see if there are free or low-cost classes being offered in your area. In the US, most Cooperative Extension offices have agricultural agents or master gardeners who offer educational programming for free.

In each chapter, you were given a recap of the main points, which was designed to serve as both an emphasis and a review. We want you to be able to use this book as not only a guide to get started, but as a reference which you can return to, if and when you need a refresher. If you're feeling comfortable with your hydroponics set-up, but want to expand into a different type of system or use a different growing medium, look no farther than coming back to Chapter 2.

If you have questions about getting set up and planted, you can always check back in with Chapter 3 and jog your memory, and if you're concerned about your daily, weekly, and monthly maintenance routines, take a quick glance back through Chapter 4. And when it's time to think about all the things you can do to bring your garden to the next level, check back in with Chapter 5. Remember, gardening is supposed to be fun and productive- if you take the time to learn about the necessary components and basic science, then you will be able to grow as big as your dreams will allow!

Of course, through all this, you know we're going to remind you about keeping that garden journal. Whether you are an artist who wants to make lists and doodles in a notebook or a software aficionado who lives to make spreadsheets, believe us when we say you will be doing yourself a great favor. Keeping track of all your activities will help you identify patterns, see what's working and what isn't, and be a more involved and active gardener.

We've come to the end of our guide, and we thank you again for taking the time to read *Gardening Secrets*. Here's to many years of healthy, happy plants! We wish you all the best in your hydroponic adventures!

Description

Have you ever wanted a garden, but you just don't have space or the soil for it? Have you thought about hydroponics, but weren't sure where to go for the information you need to get started? Hydroponic systems are self-contained, accessible to people of all physical abilities, and can provide a beautiful green oasis right in your home! There's no downside to having a hobby that is fun, educational, and productive, so why not give it a try?

These days, many people are looking to return to a healthier approach to eating and a more sustainable way of life, and if you think hydroponics is the answer for you, look no further than *Gardening Secrets*. This book can be your guide to all the knowledge you'll need to design, build, plant, and maintain the hydroponics system you've been dreaming about!

In just five easy-to-follow chapters, *Gardening Secrets* will teach you:

- the history and development of hydroponics through the centuries

- fundamental botany for plant growth

- the six basic types of hydroponic systems

- the types of growing media and their different applications

- how to create nutrient solutions for maximum plant growth

- how to choose quality seeds and seedlings for your garden

- how to choose and set-up a system that fits your needs and budget

- how to get your system running and planted

- how to observe your plants and adjust your garden accordingly

- maintenance tips and routines to keep your garden in top shape

- how to get the most out of your garden year-round

- and much, much more!

Armed with the information in *Gardening Secrets*, you'll be able to build, plant, and maintain a hydroponics system that will provide you with fresh food and flowers twelve months of the year for your health and enjoyment. You'll also learn how to troubleshoot issues with your plants and your system, recognize signs of plant distress and mechanical problems, and be encouraged to keep a garden journal to record all your activities.

You'll learn the science behind healthy plants, including what nutrients they need and how to provide those nutrients in a soil-less environment. You'll also be given insight into why plants need each of their basic requirements- water, air, light, and nutrients- and how photosynthesis works to help plants create their own fuel. When you understand the biological processes of the plant, it's much easier to understand why each step of hydroponic gardening is so important to the overall health of your plants.

Gardening Secrets is your ultimate one-stop handbook for getting on your way to the hydroponics garden of your dreams. With step-by-step guidance, easy-to-read science, and tips and routines to help you at every part of the journey, the garden you've been thinking about will become the garden you're growing in no time, so dive in, learn the basics, and get building and growing a garden you'll be able to maintain and enjoy for years. You'll be so glad you did!

www.ingramcontent.com/pod-product-compliance
Lightning Source LLC
Chambersburg PA
CBHW071535080526
44588CB00011B/1675